TECHNICAL
REPORT

Using Linear Programming to Design Samples for a Complex Survey

James H. Bigelow

Prepared for the United States Air Force

The research described in this report was sponsored by the United States Air Force under Contract FA7014-06-C-0001. Further information may be obtained from the Strategic Planning Division, Directorate of Plans, Hq USAF.

Library of Congress Cataloging-in-Publication Data

Bigelow, James H.
Using linear programming to design samples for a complex survey / James H. Bigelow.
p. cm.
Includes bibliographical references.
ISBN 978-0-8330-4163-0 (pbk. : alk. paper)
1. United States. Air Force—Public opinion. 2. Linear programming. 3. Social surveys—United States—Case studies. I. Title.

UG633.B54 2007
358.4'1330723—dc22

2007031268

Published 2007 by the RAND Corporation
1776 Main Street, P.O. Box 2138, Santa Monica, CA 90407-2138
1200 South Hayes Street, Arlington, VA 22202-5050
4570 Fifth Avenue, Suite 600, Pittsburgh, PA 15213-2665
RAND URL: http://www.rand.org/
To order RAND documents or to obtain additional information, contact
Distribution Services: Telephone: (310) 451-7002;
Fax: (310) 451-6915; Email: order@rand.org

Preface

This document was prepared within RAND Project AIR FORCE's Manpower, Personnel, and Training Program. The research was performed as part of a fiscal year 2006 project, "Enhancing the Common Airman Culture," sponsored by the Air Force Deputy Chief of Staff for Personnel, Headquarters U.S. Air Force (AF/A1).

The objective of this project was to recommend strategies to strengthen the overarching airman culture in order to reinforce a common set of core values and appropriate responses to violations of those values. One component of this research was an online survey of Air Force personnel.

This document describes the method we developed to design the sample of Air Force personnel to invite to participate in the survey. It is intended for readers with an interest and expertise in quantitative survey methods and/or statistics and survey sampling.

RAND Project AIR FORCE

RAND Project AIR FORCE (PAF), a division of the RAND Corporation, is the Air Force's federally funded research and development center for studies and analyses. PAF provides the Air Force with independent analyses of policy alternatives affecting the development, employment, combat readiness, and support of current and future aerospace forces. Research is conducted in four programs: Aerospace Force Development; Manpower, Personnel, and Training; Resource Management; and Strategy and Doctrine.

Additional information about PAF is available on our Web site:
http://www.rand.org/paf/

Contents

Figures and Tables

Figures

Tables

Summary

This report describes a method we developed to design and select the sample of Air Force personnel who would be asked to participate in a survey on cultural attitudes. The survey is long and complex and has many competing goals, so the design problem has no simple answer. In the simplest surveys, one seeks only to estimate a population parameter—say, the proportion of the population that prefers chocolate to vanilla—from the responses of the people in the sample. But, in this case, the design considered the following:

- We intended to build multiple three- and four-way crosstabs from the responses (e.g., by grade and AFSC family,[1] and by grade and gender), so we needed to select a sample from which we could estimate differences in population parameters in different three- or four-way crosses with the desired precision.
- Because the survey was very long, we split the sample into three sections and asked the people assigned to each section to respond to only part of the survey. Thus, we needed to trade off sample size (and therefore precision) in one section versus the others.
- Finally, our survey (which we refer to as the CULTURE survey) was fielded shortly after another survey (the HEALTH survey), and because we feared that survey fatigue would reduce the response rate among people in both samples, we wanted to design a sample for our survey that overlapped as little as reasonably possible with the other survey's sample.

Our method partitioned the population into cells. By the definition of a partition, each member of the population belonged to a unique cell. We formed the cells by taking the eight characteristics needed to define all the three- and four-way crosstabs mentioned above and using them to define one gigantic eight-way cross classification. There were over a million cells. More than 98 percent of them were empty, and almost half the nonempty cells contained only one person. For example, there just are not very many company-grade female, Hispanic, Roman Catholic pilots in the Air National Guard who are assigned to the Air Combat Command. (Our method ignores the empty cells, since they contain no personnel to select for our sample.)

We recovered a particular three- or four-way crosstab from the cells (i.e., the eight-way classification) by summing over characteristics not used to define that crosstab. Aggregate data

[1] The Air Force Specialty Code (AFSC) is a code for the skills an individual possesses.

obtained in this way are called *marginal* data, and we will refer to each three- or four-way cross as a *marginal cross*.

Our method used linear programming, a well-known procedure for optimizing (maximizing or minimizing) a linear function subject to linear constraints. The primary variables in our linear program were the expected number of people drawn from each cell for assignment to each section of the survey. (For large cells, there was no practical difference between the number drawn and the expected number. For a cell containing only one person, the expected number was the probability that the person was in the sample.) The constraints ensured that

- the number of people drawn from a cell could not exceed the cell population
- the sample taken from each marginal cross was at least as large as a specified quantity
- the sample taken from each cell was at least as large as another specified quantity. This quantity was selected to ensure that an adequate number of people from the cell were in the sample for every marginal cross to which the cell contributed.

The last two constraint types ensured that estimates of population parameters would have, as far as possible, the desired precision.

Merely because our method used linear programming—which solves an optimization problem—did not mean that we regarded the sample it designed to be "optimal" in a real-world sense. The CULTURE and HEALTH surveys each had its own desired precision, as did each of the sections within the CULTURE survey, and we could only improve the precision of one survey/section at the expense of another.[2] So we embedded a number of "design levers" in the linear program. By manipulating them, we caused the method to generate sample designs that gave different priorities to the various surveys/sections. Flesh-and-blood humans examined the various designs and, ultimately, picked one.

To aid in this process, we defined a number of summary measures of design quality. An obvious one was total sample size. Others were counts of "problem crosses," i.e., marginal crosses for which the sample did not satisfy the last two constraint types and hence would not yield the desired precision. By varying the threshold at which we considered a cross to be a problem, we could construct a whole family of measures.

It seemed to the survey team that the HEALTH survey had rather little precision unless it received a very high priority, but this caused the CULTURE survey to lose too much precision. This observation persuaded the survey team to investigate the effect of allowing the two samples to overlap. Indeed, giving the HEALTH survey a high priority causes the CULTURE survey to lose much less precision in the samples with overlap than in those without.

The last step in the process was to select an actual sample. The linear program solution that we ultimately chose gave us the expected number of people to draw from each cell for each section. We developed a variant of systematic sampling to choose the actual individuals to include in the sample. In our method, each member of the population had some chance of

[2] The CULTURE survey was partitioned into three sections, so assigning a person to a section also assigned that person to the CULTURE survey; anyone assigned to the CULTURE survey had to be in one of the three sections. The HEALTH survey was not divided into sections, so we do not use the term "section" when referring to the HEALTH survey. A person assigned to the overlap was assigned to both the HEALTH survey and a section within the CULTURE survey.

being assigned to any survey/section. The sample our method selected matched the linear programming solution quite closely. Both the overall sample and the sample taken from each large cell had sizes close to the expected sizes calculated by the linear program.

There is no guarantee, however, that the precision of estimates we calculate when we analyze the survey responses will equal the precision we sought when we designed the sample. The actual precision will depend on factors that were unknown when we designed the sample (for example, the actual response rate). Moreover, we could only design the sample to ensure adequate precision of estimates we anticipated we were going to make. Once we begin analyzing the survey responses, we may discover that some quantities we did not anticipate estimating are much more interesting.

In short, we will inevitably judge the adequacy of the sample using standards that we do not fully know at the time we design the sample. We can only guess what those standards will be, design the sample to our guess, and let the chips fall where they may.

Acknowledgments

This report would not have been possible without contributions from many others. Discussions with Chaitra Hardison (RAND) ensured that the methodology developed in this work addressed the needs of the project. John Crown (RAND) prepared extracts of Air Force personnel data and consulted with the author on data issues. The author thanks them both.

Edward Chan (RAND) and James R. Chromy (Senior Fellow at RTI International) reviewed a draft of the report. Their helpful and insightful comments led to substantial improvements in the report. The author, of course, remains responsible for any remaining errors or shortcomings.

Abbreviations

ACC	Air Combat Command
AF/A1	Air Force Deputy Chief of Staff for Personnel, Headquarters U.S. Air Force
AFMA	Air Force Manpower Agency
AFMC	Air Force Materiel Command
AFSC	Air Force Specialty Code
AFSOC	Air Force Special Operations Command
AMC	Air Mobility Command
CID	cell identification
fpc	finite population correction
GAMS	General Algebraic Modeling System
IRR	Individual Ready Reserve
NCO	noncommissioned officer
PAF	Project AIR FORCE
PID	personal identification

CHAPTER ONE

Introduction

In the summer of 2005, a RAND Corporation study team was asked to assist the Air Force in assessing Air Force culture and its relationship to a range of behaviors it deemed aberrant. We developed a questionnaire for a survey of Air Force personnel on cultural attitudes (henceforth called the CULTURE survey), and designed a sample of the population to receive email invitations to participate in the survey. The design needed to meet a number of goals that may concern other survey researchers as well: (1) minimize the number of people asked to participate so as to reduce the survey burden on a population already frequently invited to take surveys; (2) reflect response rates we could anticipate from previous surveys of the population; (3) ensure adequate representation of a number of minorities of interest (rank, job type, race and ethnicity, gender, religion, and component[1]); (4) sample enough people in each of the overlapping subset categories of interest (e.g., black female noncommissioned officers [NCOs]) to allow for statistically meaningful comparisons; and (5) minimize (to zero, if possible) the number of service members invited to take both this survey and another survey (the HEALTH survey) on an overlapping set of topics scheduled for the same time period.

We describe here the method we developed for designing joint samples for the CULTURE and HEALTH surveys. The Air Force personnel inventory consists of approximately 350,000 active, 105,000 Air National Guard, 75,000 Air Force Reserve, and 150,000 civilian personnel. While our survey drew samples from all these groups, in this report we illustrate the method for the Guard and Reserve only. We wish to select a sample of these individuals that is large and diverse enough to allow us to draw conclusions about how their attitudes are related to their various personal and professional characteristics.

The many competing goals of this survey preclude there being a sample design that is best in every way—i.e., an optimal design (such a point was made in Adams et al., 2003). Rather, the design must strike a balance among the many goals. But even though there is no optimal design, some designs are better than others. Our methodology generates relatively good samples for flesh-and-blood humans to examine. The humans must exercise their judgment in deciding which sample design to choose.

Next, we describe the factors that entered into the design of the sample. This is perhaps most easily done if we begin with a very simple survey and successively add layers of complexity to it.

---

[1] By *component,* we mean Active Duty, Air National Guard, Air Force Reserve, or Air Force civilians.

Step 0: A Simple Survey

The simplest surveys—the ones described in the early chapters of textbooks on sampling (e.g., Cochran, 1977; Kish, 1995)—seek to estimate the average value of a variable Y in a population of size N from data on that variable measured for a sample. For example, one might wish to determine the fraction of a population that prefers vanilla ice cream to chocolate from the answers given by a sample from that population. It makes sense to design a sample that is smallest for a given precision or, if the cost of collecting data differs for different members of the population, to design a sample that is cheapest for a given precision. We would call this an *optimal* sample.

Step 1: A Survey to Support Multiple Crosstabs

The next step in complexity occurs if we want to estimate the average value of the variable Y, not in the entire population but in specified subsets of the population. If the subsets are mutually disjoint, the problem reduces to multiple simple surveys. But if the subsets overlap, the data collected on each member of the population may contribute to the average value of Y in several subsets.

In our case, the subsets are the crosses in three- and four-way crosstabs defined by known characteristics of members of the population. For example, we wished to tabulate responses of Air Force Reserve personnel by grade and AFSC family,[2] by grade and major command, by grade and gender, and by grade and race. Each respondent had a grade, an AFSC family, a major command, a gender, and a race, and data from each respondent therefore contribute to one cross in every table.

If we are willing to specify a desired precision for each cross, we can still design a sample that is smallest or cheapest, and that we might still, therefore, be willing to call "optimal." Chromy (1987) discusses exactly this problem.

Step 2: A Survey with Multiple Sections

Next, we split the survey into sections. The CULTURE survey was so long that we feared many would-be respondents would quit in the middle—or even not start at all—if they were asked to complete every question. So each respondent was assigned to one of three sections (imaginatively called A, B, and C). We asked each respondent to answer a group of questions that constituted the "core" of the survey, and a further set of questions specific to the section he or she was assigned to.

Sections A, B, and C are in direct competition with one another, in that assigning a person to one section precludes his or her being in another. Assigning him or her to section A therefore sacrifices precision in sections B and C. If there are enough people in the population

[2] The Air Force Specialty Code (AFSC) is a code for the skills an individual possesses.

to simultaneously achieve the desired precision in all the affected crosses, there is no problem. But if there are not enough people, we must trade off precision in one section for precision in another. We can no longer call any particular sample optimal.

Step 3: Multiple Competing Surveys

Finally, a second survey of Air Force personnel was being fielded by the Air Force Surgeon General to assess the health of Air Force personnel (henceforth called the HEALTH survey). It cross-tabulated personnel on different characteristics. We had the opportunity to design samples for both surveys, and we sought to design samples that would balance the competing desires of the two survey groups. As with the competing sections within our own survey, more precision in one survey generally implies less precision in the other. Thus, no particular pair of samples should be considered optimal.

We posited survey fatigue as the mechanism by which the HEALTH survey, which was fielded first, could affect the CULTURE survey. An individual in only one sample would respond with one probability (nominally 50 percent).[3] We assumed that an individual in both samples would respond with the same probability to the HEALTH survey but with a lower probability (nominally 25 percent) to the CULTURE survey.[4]

Multiple Crosstabs Require Thousands of Small Cells

As mentioned, we wished to tabulate responses of Air Force personnel by various combinations of personal and professional characteristics. Earlier we listed five of the characteristics we considered: grade (six values), AFSC family (nine values), major command (six values), gender (two values), and race (five values). There are $6 \times 9 \times 6 \times 2 \times 5 = 3{,}240$ possible combinations of values for these five characteristics alone, and adding the remaining characteristics we considered (component, faith, and installation) expands the number to over a million. In fact, there were almost 20,000 different combinations of characteristics possessed by at least one person in the Air National Guard or Air Force Reserve. Each of these nonempty combinations defines a cell from which we sample.

The sheer number of cells poses neither a conceptual nor a computational problem. Standard statistical formulas remain valid for any number of cells. The software we use easily accommodates problems much larger than ours.

However, many of our cells are small. The average cell contains only nine people, and over 8,000 cells contain only one person. (Most of the approximately 180,000 Reservists and Guard members, of course, are in a relatively few large cells.)

---

[3] There was a 50-percent participation rate in the 2003 Air Force Climate Survey. This survey, which all Air Force personnel were urged to take, measured how people felt about leadership, supervision, training, recognition, and other aspects of the Air Force (personal communication from Air Force Manpower Agency [AFMA], which conducted the survey).

[4] The 25-percent response rate for people invited to take both surveys is conjecture because we had no data to base it on.

The presence of small cells makes it impractical to draw a stratified random sample—i.e., a simple random sample of predetermined size from each cell. Instead, we employ a sampling method (described in Chapter Six) that has a nonzero probability of assigning each member of the population to every survey/section. The probability of a given assignment is the same for all members in the same cell, but it generally differs for different cells. The method ensures that the size of the sample taken from a large cell is close to the expected sample size for that cell, so if there were no small cells, we would very nearly obtain a stratified random sample. The method also ensures that the overall size of the sample is very close to the expected size.

The proper formulas for analyzing the survey responses depend on the sample selection method. Horvitz and Thompson (1952) present general formulas for samples that, like ours, are drawn without replacement from finite populations, with unequal but positive selection probabilities.

Cells and Crosses

We use the terms *cell* and *cross* (or *marginal cross*) throughout this report, so it is worthwhile to describe them and the relationship between them more fully. Each individual in the population has values for each of eight personal or professional characteristics—component, grade, major command, AFSC family, race, gender, faith, and installation. A cell consists of all people who have the same values for all eight of these characteristics. Since no person has two different values for any characteristic, the cells are disjoint. A (marginal) cross, on the other hand, consists of all people who have the same values for only some selected characteristics but may have different values for the remaining characteristics. Thus a cross defined by component = Air National Guard, grade = Airman, and AFSC family = Logistics will contain all people who share those values of component, grade, and AFSC family, but it will include people with different genders (both female and male), different faiths (Protestant and Roman Catholic, as well as other faiths), and so on. It is evident that a cross is a union of cells, and the data in the cross are obtained by summing over those cells. Data obtained in this way are generally referred to as marginal data; hence our name, *marginal cross*.

Outline of the Method

Our method is based on linear programming, a technique for solving problems of the following form (e.g., see Hillier and Lieberman, 2005): Given constants a_{ij}, b_i, and c_j for $i = 1,\ldots,m$ and $j = 1,\ldots,n$, find values for the variables x_j that

$$\begin{aligned} &\textit{Minimize} && \sum_{j=1}^{n} c_j x_j \\ &s.t. && \sum_{j=1}^{n} a_{ij} x_j \le b_i \quad i = 1,\ldots,m \\ & && x_j \ge 0 \qquad\quad j = 1,\ldots,n\ . \end{aligned} \tag{1.1}$$

In our method, each variable x_j corresponds to the expected number of people from each cell invited to take either each section of the CULTURE survey, the HEALTH survey, or both HEALTH and a section of CULTURE. Some of the constraints ensure that no cell contributes more responses than possible, taking its population and response rates into account. Other constraints ensure that the "core" sample for CULTURE equals the union of our A, B, and C samples. Still other constraints ensure that, to the degree possible, each marginal cross we wish to examine contains a sample that is large and diverse enough to provide the desired precision.

Others have suggested formulating sample design problems as optimization problems. Chromy (1987) addressed much the same problem we do, but formulated it with nonlinear constraints. This is perhaps more natural, since Chromy's primary constraint was intended to restrict the size of the standard error of quantities estimated from survey results, and the standard error of an estimate is not linear in the size of the samples drawn from the various cells.[5] But no software package was readily available to us that would solve a large nonlinear optimization problem, and solving such problems is problematic in any case. By formulating the problem with linear constraints, we make it possible to use a software package called the General Algebraic Modeling System (GAMS), which both provides a means to describe the problem algebraically and reliably solves very large linear programs.

Lu and Sitter (2002) proposed a method for designing survey samples that uses linear programming. In effect, they address a problem with a single cross to which all cells belong but which is otherwise very similar to ours. They specify the overall size of the sample as an input, and their method guarantees that their sample will have the desired size; we specify the desired sample size for each marginal cross, but it may not be possible to achieve the desired size for every cross. They insist that the sample drawn from each cell should be as nearly as possible proportional to the size of the cell; we impose similar proportionality constraints for each marginal cross, which also may not be achievable. The main difference is that they calculate a probability distribution over all admissible allocations of the sample to the cells, whereas we calculate only the expected allocation. The resulting combinatorial explosion leads Lu and Sitter to a much larger linear program than we would obtain for the same problem. Indeed, Lu and Sitter (2002, p. 200) "give some examples with from 80 to 300 stratification cells to illustrate the ability of the new methodology to handle large problems."

[5] Chromy uses the variance formula for stratified random sampling, which is inappropriate for our problem. But Chromy's problem could be reformulated to use a Horvitz-Thompson formula.

To design a sample, we solve two linear programs in sequence. Because the response rates are less than 100 percent and because the two surveys and the sections of the CULTURE survey compete with one another for respondents, it is not possible to meet all the constraints on precision. In the first problem, we minimize the total number of additional people we would need to achieve the desired sample size for every cross and simultaneously meet the proportionality constraints, summed over all marginal crosses. For the second problem, we constrain the shortfall for each marginal cross to the minimum value computed in the first problem, and we minimize the total size of the samples, counting people twice if they are in the overlap.[6]

Even though we solve an optimization problem to obtain the sample, we do not consider it to be "optimal" in the real-world sense. We have included a handful of parameters in the model—call them *design levers*—that do not have obvious "best" values, and by varying these levers we can generate a wide range of samples for flesh-and-blood humans to evaluate. One lever, for example, affects the lower bound we place on the precision of estimates the sample in each marginal cross can support. Another governs how the population in a cell will be divided between the two surveys, if both cannot have all that the surveyors want.

The solution to the two linear programs specifies the expected number of people from each cell who should be in our sample. It specifies how many should be assigned to the CULTURE survey, the HEALTH survey, or both, and how many in the CULTURE sample should be assigned to each section. Now we must construct a list of individuals in the Air Force that assigns approximately the specified numbers of people from each cell to each survey/section as specified in that design.

We used a variant of systematic sampling (e.g., see Cochran, 1977, Chapter 8; or Kish, 1995, Chapter 4). In the standard version of systematic sampling, one takes every *k*th individual of the population after a random start. In our problem, however, individuals of the population are not simply included or excluded; rather, they are either excluded or assigned to a survey/section. Moreover, the probability for each assignment is not the same for all individuals (although the probability is the same for individuals in the same cell). Our variant of systematic sampling accommodates these complexities.

Plan of the Report

In Chapter Two, we discuss the data in the Air Force personnel file from which we selected our sample. Chapter Three presents constraints on the numbers of individuals who can be in each section of our sample—both in the other survey sample and in the overlap. Basically, these constraints ensure that the number of people sampled from each cell is no greater than the population of that cell.

Chapter Four formulates the constraints that govern precision. One kind of precision constraint requires the sample from each marginal cross to exceed a specified threshold. The other kind requires the sample from each cell in the marginal cross to exceed a specified pro-

[6] The method is easily modified to find the sample that yields the greatest precision given constraints on sample size. One merely imposes constraints on sample size in the first problem. This eliminates the need to solve the second problem.

portion of the cell's population. Together, these constraints ensure that the sample from each marginal cross is large and diverse enough to yield reasonably precise estimates for the entire population of the cross.

In Chapter Five, we report a range of samples designed by the method and discuss tradeoffs. As mentioned earlier, the several surveys/sections compete for participants. If the terms of that competition favor one survey/section, its sample size will increase, and estimates made from its sample will have greater precision (smaller variances). But estimates made for marginal crosses in the other surveys/sections will be less precise.

In Chapter Six, we describe how we selected the sample of individuals from the Air Force personnel file who will be asked to take the survey. We use the solution generated by our method to specify the probabilities that each individual will be included in the sample asked to take each survey/section.

CHAPTER TWO

Personnel Data

Our primary source of data is the Air Force personnel file, a database maintained by the Air Force that describes the history and current status of all its military personnel. The linear programs we formulate require personnel inventories obtained from the personnel file, by cell. Once we have designed a sample—i.e., determined how many individuals from each cell we will ask to take a survey and which survey and section to assign them to—we select specific individuals from the personnel file in numbers that approximately match that design.

We define the cells using eight personal and professional characteristics of each individual. All eight characteristics correspond to data elements in the Air Force personnel file. Table 2.1 shows the Guard and Reserve personnel characteristics used by the two surveys and the categories we have defined for them. As discussed in Chapter Four, neither survey made use of all eight categories (CULTURE used seven in its three- and four-way crosstabs; HEALTH used three).

CULTURE and HEALTH did not survey the same populations. The personnel in CULTURE but not HEALTH included individuals with a grade of "General Officer" and all members of the Air National Guard. CULTURE included people at more installations than did HEALTH. The personnel in HEALTH but not CULTURE were mostly members of the Individual Ready Reserve (IRR). Thus, we constructed three arrays of personnel inventories, as shown in Table 2.2. These arrays are the primary data sources for our linear programs.

One feature to note is the large number of small cells. The average nonempty cell contains about nine people, and over 8,000 cells have only one person. It seems excessive to have so many cells and such small ones. But we are cross-classifying on only eight characteristics. Except for installations, we have nine or fewer categories per characteristic. Neither eight characteristics nor nine categories per characteristic seem unreasonable. The large number of cells is an inevitable consequence of choosing these characteristics and their categories. The Air Force personnel file tells us how many people each cell contains, so the fact that there are so many small cells is also unavoidable. There just are not very many company-grade female Hispanic Roman Catholic pilots in the Air National Guard who are assigned to Air Combat Command (ACC).

Table 2.1
Categories That Define Cells for Guard and Reserve Personnel

Characteristic	Categories
Component	Air National Guard Air Force Reserve
Grade	Airman Company grade Field grade General Officer NCO Senior NCO
Major command	ACC Air Force Materiel Command (AFMC) Air Force Special Operations Command (AFSOC) Air Mobility Command (AMC) Other None
AFSC family	Acquisition Logistics Medical Office of Special Investigations Other Other operations Pilot Professional Support
Race	Black, non-Hispanic Hispanic or Latino Other Unknown White
Gender	Female Male
Faith	Evangelical Non-Christian Protestant Roman Catholic Unknown
Installation	38 installations plus "Other"

Table 2.2
Arrays of Guard and Reserve Personnel Inventories

Eligible for . . .	Number of Nonempty Cells	Number of Personnel
CULTURE only	7,436	120,338
HEALTH only	183	9,207
CULTURE and HEALTH	11,924	46,584

As mentioned earlier, the presence of small cells makes it impractical to draw a stratified random sample. Because stratified random sampling is so well understood, it might seem attractive to merge the small cells into a small number of large strata and to draw a stratified random sample from the restratified population. But if one simply merges all small cells into a single stratum, one may lose the ability to oversample members of the population with relatively rare characteristics. Avoiding this outcome, or even determining whether it has occurred, poses its own difficulties. Moreover, it rankles to allow the tool to dictate the problem.

CHAPTER THREE

Constraints on Numbers of People Assigned to Surveys and Their Sections

In this chapter and the next, we present the constraints that appear in our linear program. This chapter presents constraints that ensure that the sample taken from a cell is no larger than the entire population of that cell. These constraints also govern the assignment of people in the sample to the CULTURE survey or HEALTH survey or both, and to sections within the CULTURE survey. The next chapter presents the constraints that ensure the precision of the estimates we make from survey responses.

Variables

The primary variables in the linear program represent the numbers of participants assigned to each survey and, for the CULTURE survey, to each section.

Sample(*Assign*,*Cell*)	=	Number of people in the cell asked to take the survey/section indicated by the assignment. For assignments to the CULTURE survey, this is limited to people assigned to CULTURE only. It does not include people assigned to both CULTURE and HEALTH.
Overlap(*Assign*,*Cell*)	=	Number of people in the cell asked to take both the HEALTH survey and one of the sections of the CULTURE survey.

The variables *Sample* and *Overlap* are allowed to take on fractional values and therefore cannot literally be numbers of people. Instead, we interpret them as expected numbers of people.[1] Then the ratio of *Sample*(*Assign*,*Cell*) or *Overlap*(*Assign*,*Cell*) to the size of *Cell* is the probability that a given individual in *Cell* will have the indicated assignment.

The assignment to survey and section, denoted by *Assign*, can take on the values shown in Table 3.1.

[1] In Chapter Six, we use the values of *Sample* and *Overlap* as the basis for selecting actual samples—i.e., lists of individuals—for the CULTURE and HEALTH surveys. We can imagine applying our method repeatedly to generate a population of samples and taking expectations over that population.

Table 3.1
Possible Values of *Assign*

Value	Interpretation
Core	Assigned to CULTURE survey, any section
A	Assigned to CULTURE survey, section A
BC	Assigned to CULTURE survey, section B or C
HEALTH	Assigned to HEALTH survey

Observe two things about this list. First, the assignments are not mutually exclusive. A person assigned to the CULTURE survey will be assigned to one of its sections, and therefore two values of *Assign* will apply. Second, sections B and C of the CULTURE survey do not have separate assignments. This is because responses to sections B and C are tabulated in the same way (see Chapter Four), and therefore we assign the same number of people from each cell to the two sections.

Constraints on Assignments

We define the following:

Pop(*Cell*) = Number of people in this cell.

CULTURE(*Cell*) = "T" (for "True") if people in this cell are eligible for the CULTURE survey;
"F" (for "False") if they are not eligible.

HEALTH(*Cell*) = "T" if people in this cell are eligible for the HEALTH survey;
"F" if they are not eligible.

There are four assignment constraints for each cell. The first limits the total number of people assigned from a cell to no more than the total number of people in that cell.

$$Sample(\text{“}Core\text{”}, Cell) + Sample(\text{“}HEALTH\text{”}, Cell) \leq Pop(Cell). \qquad (3.1)$$

The second constraint requires that people assigned to any section of CULTURE are also assigned to the core survey.

$$Sample(\text{“}Core\text{”}, Cell) = Sample(\text{“}A\text{”}, Cell) + 2 \times Sample(\text{“}BC\text{”}, Cell). \qquad (3.2)$$

In addition, people cannot be assigned to a survey if they are not eligible for it.

$$Sample(\text{“}Core\text{”}, Cell) = 0 \quad if \;\; CULTURE(Cell) = \text{“}F\text{”}. \qquad (3.3)$$

$$Sample(\text{“}HEALTH\text{”}, Cell) = 0 \quad if \;\; HEALTH(Cell) = \text{“}F\text{”}. \tag{3.4}$$

Equations (3.5)–(3.8) place similar limitations on the people in the overlap between the CULTURE and HEALTH samples:

$$Overlap(\text{“}Core\text{”}, Cell) \leq Sample(\text{“}HEALTH\text{”}, Cell). \tag{3.5}$$

$$Overlap(\text{“}Core\text{”}, Cell) = Overlap(\text{“}A\text{”}, Cell) + 2 \times Overlap(\text{“}BC\text{”}, Cell). \tag{3.6}$$

$$Overlap(\text{“}Core\text{”}, Cell) = 0 \quad if \;\; CULTURE(Cell) = \text{“}F\text{”}. \tag{3.7}$$

$$Overlap(\text{“}HEALTH\text{”}, Cell) = 0 \quad if \;\; HEALTH(Cell) = \text{“}F\text{”}. \tag{3.8}$$

Equations (3.3), (3.4), (3.7), and (3.8) do not actually appear in the linear program. Instead, linear programming software algebraically eliminates these variables from the problem before solving it. Because Eqs. (3.1), (3.2), (3.5), and (3.6) are replicated for each cell, they contribute thousands of constraints to each linear programming problem. The Guard and Reserve inventory has about 20,000 cells (see Table 2.2). So Eqs. (3.1), (3.2), (3.5), and (3.6) contribute about 4 × 20,000 constraints to the Guard/Reserve problems, respectively. These are large linear programs.

CHAPTER FOUR

Constraints to Ensure Adequate Precision of Estimates

This chapter presents the constraints on the sizes and compositions of the samples taken from the various marginal crosses. These constraints are intended to ensure adequate precision, as far as possible, for estimates of population parameters made for each cross.

Table 4.1 identifies all the three- and four-way crosstabs for which we planned to make estimates. Taking the first line, we partition all personnel who answer the core questions into categories (crosses) whose members have the same component (e.g., Air Force Reserve), the same grade (e.g., Airman), and the same AFSC family (e.g., Acquisition).

Note that every person in a given cross is in the same component. This is what allowed us to design the Reserve and Guard personnel sample separately from the Active Duty sample.

A considerable number of crosses have rather small personnel inventories. Of 905 crosses for Guard and Reserve personnel, 208 crosses contain fewer than 100 people; 121, fewer than 50; and 54, fewer than 10. The very small Guard and Reserve crosses are crosses for General

Table 4.1
Marginal Crosses for Which We Make Population Estimates

		Data Elements Held Constant			
Assignment	Number of Nonempty Crosses	1	2	3	4
Core	86	Component	Grade	AFSC family	
Core	71	Component	Grade	Major command	
Core	24	Component	Grade	Gender	
Core	59	Component	Grade	Race	
Core	20	Component	Race	Gender	
Core	168	Component	AFSC family	Race	Gender
A	86	Component	Grade	AFSC family	
A	71	Component	Grade	Major command	
BC	20	Component	Gender	Faith	
BC	60	Component	Grade	Faith	
BC	50	Component	Race	Faith	
HEALTH	190	Component	Grade	Installation	

Officers and crosses for pilots or special investigators who are female. There are only a few hundred in each of these categories in the entire Air Force, so it is not surprising that crosses containing subsets of these categories are small.

Formulas for Precision

The survey will provide responses to various questions by a sample of people from each marginal cross, and we wish to estimate from that sample how the entire population of that cross would have answered the same questions. For example, the survey could ask each respondent whether he or she preferred vanilla ice cream to chocolate. From their responses, we wish to estimate the proportion of the entire population of the cross that would have said they preferred vanilla if we had been able to perform a complete census.

Because we have only a sample rather than a census, our estimate is subject to error, usually measured by the variance of the estimate. We want to design the sample so the variances will be "small enough." Generally, we will consider the difference between two different marginal crosses to be meaningful (that is, large enough to care about) if it is larger than some threshold. For example, we might consider it meaningful if vanilla is preferred to chocolate by as many as 5 percent more of the entire population of one cross than another. We want to design our sample so that meaningful differences are statistically significant.

The proper formulas for the variances depend on the sample selection method. Horvitz and Thompson (1952) presented a general variance formula for samples drawn from finite populations without replacement and with (possibly) unequal selection probabilities. The explicit variance formulas for simple random samples and stratified random samples without replacement that one sees in standard textbooks (e.g., Cochran, 1977; Kish, 1995) can be obtained as special cases of the Horvitz-Thompson formula. The Horvitz-Thompson formula contains the selection probabilities for pairs of individuals as well as the individual selection probabilities, and the pairwise probabilities are often difficult to work out. Some approximations are compared in Stehman and Overton (1994).

All variance formulas, however, are functions of the sample design variables—*Sample*(*Assign*,*Cell*) and *Overlap*(*Assign*,*Cell*), in our case—so imposing a limit on the size of the variance thus implicitly constrains the sample design variables. Unfortunately, these variance functions are not linear, so they cannot be used directly as constraints in a linear program. Software for nonlinear programming problems is less available and less robust than software for linear programming problems, so there is a considerable practical advantage to formulating linear versions of the variance constraints.

We base our linear precision constraints on two features common to all the variance formulas.[1] First, the variance decreases as the sample size increases (often, the variance is inversely

[1] This implies that our precision constraints can be tuned to reflect any variance formula and therefore any method for selecting the sample.

proportional to sample size). Second, for equal sample sizes, the variance is smaller when the sample more nearly mirrors the cross's population. To capture the first feature, we constrain the size of the sample from a cross to exceed a specified threshold. To capture the second feature, we insist that the sample from each cell in the cross exceed a specified proportion of the cell's population.

Admittedly, these linear constraints only roughly simulate the effects of the nonlinear constraints they replace. But the nonlinear constraints only roughly produce the effects desired of them—namely, that the variances of estimates made from the survey responses will be "small enough." The variances depend on more than the sample design. They also depend on

- the response rate (the fraction of those invited to take the survey who actually do so)
- the possible answers to a survey question (e.g., yes or no, pick a number from 1 to 7, continuously variable answers)
- the distribution of answers actually given (e.g., if almost all respondents in a cross give the same answer to a question, the estimate for the entire population of the cross will have a very small variance).

Some of these factors cannot be known until after the survey has been fielded and the responses collected, so one cannot specify the constraints on survey design that will yield exactly the desired variances. Usually, they are chosen to ensure that the variances will be no larger than desired, but then they will tend to constrain the sample design more severely than necessary. Thus, whether one uses the linear or nonlinear constraints, they are always approximate.

Constraints on Sample Size by Marginal Cross

Constraints based on the first feature require that the number of people from a marginal cross who respond to the survey should, if possible, equal or exceed a specified size. There are three new elements to consider.

Response Rates

First, we are interested in the number of responses, not the number assigned. We define

$SRR(Assign)$ = Response rates for *Sample* variables.

$ORR(Assign)$ = Response rates for *Overlap* variables.

Table 4.2 shows the response rates that we assumed.

Table 4.2
Assumed Response Rates

Variable	Assignment	Survey	Response Rate
Sample	Core	CULTURE	0.5
Sample	A	CULTURE	0.5
Sample	BC	CULTURE	0.5
Sample	HEALTH	HEALTH	0.5
Overlap	Core	CULTURE	0.25
Overlap	A	CULTURE	0.25
Overlap	BC	CULTURE	0.25

Desired Responses per Marginal Cross

Second, we need to specify the desired number of responses for each cross. We define

Rqmt(*Cross*) = Required responses in this cross to achieve the desired precision.

The HEALTH survey team provided us with the numbers of people in each cross that they wanted to participate in their survey, and we simply multiplied those numbers by the assumed response rate.

The CULTURE survey team decided that a sample of *NomCross* = 550 responses from a very large (effectively infinite) cross would provide adequate precision. We then applied a finite population correction (fpc) that reduced the desired responses. The formula is

$$\left(\frac{1}{n}-\frac{1}{N}\right)=\frac{1}{NomCross}, \tag{4.1}$$

where n is the number of responses needed from a cross of size N to give the same precision as a simple random sample of size *NomCross* taken from a very large cross.[2] Of course, the number of responses n cannot be larger than the cross size N multiplied by the response rate (at most, 0.5). Thus for any cross smaller than *NomCross*, we can invite the entire population of the cross to participate and we still will not have the desired number of responses.[3]

---

[2] This will be true if the people in a cross who respond to the survey constitute a random sample of the population of the cross.

[3] A reviewer questioned whether our use of the fpc was appropriate. The Air Force personnel file changes over time as people enter and leave the Air Force, change jobs, receive promotions, and so on. It is correct to use the fpc only if we wish to make inferences about a snapshot of the Air Force population. The reviewer suggested that instead we should be making inferences about future snapshots as well as the current one (i.e., a superpopulation). The superpopulation is theoretically infinite, and therefore we should not be using an fpc. Others have made similar arguments (see Elliott, Zaslavsky, and Cleary, 2006). We will not comment on the merits of this suggestion, since our use of the fpc is a matter of historical fact and cannot be undone. Instead, we present an estimate of the difference it would have made if we had not used the fpc. In the case we selected as our final sample design, the number of responses calculated by our method was strictly between the required responses calculated with and without the fpc for only 62 out of 703 marginal crosses. Dropping the fpc should affect results for these crosses only. We calculated a total requirement for these 62 crosses of 25,000 responses with the fpc and 31,000 responses without. Our method calculated 27,000 actual responses. In the remaining 641 crosses, the number

"If Possible"

The third new element is that we may not be able to assign enough people to a survey/section to obtain the desired number of responses. To allow for fewer responses than desired, we introduce additional variables:

Short(*Cross*) = The amount by which responses in this cross fall short of the desired number.

Constraints on Sample Size by Cross

We can now write the constraint on the overall size of the sample for each cross. We define

Assign(*Cross*) = Assignment people must have to be in this cross's population.
Contains(*Cross*,*Cell*) = 1 if *Cross* contains the people in *Cell*; 0 otherwise.

Then,

$$\begin{aligned}&\sum_{Cell}\left[Contains(Cross,Cell)\times Sample\left(Assign(Cross),Cell\right)\times SRR\left(Assign(Cross)\right)\right]\\&+\sum_{Cell}\left[Contains(Cross,Cell)\times Overlap\left(Assign(Cross),Cell\right)\times ORR\left(Assign(Cross)\right)\right]\\&\geq Rqmt(Cross)-Short(Cross).\end{aligned}\tag{4.2}$$

Note the appearance of *Assign*(*Cross*), rather than the unmodified *Assign*, to specify which elements from the arrays *Sample*, *Overlap*, *SRR*, and *ORR* to include in the constraint. This indicates that wherever a value of *Assign* is needed in the expression, it should be the value associated with *Cross*.

Of course, we want to keep the *Short*(*Cross*) variables as small as possible. Our first step in designing a sample, therefore, is to solve a linear program with the objective of minimizing the sum over all crosses of *Short*(*Cross*).

Constraints on Sample Size by Cell

A sample from a cross is called *proportionate* if it includes the same fraction of people from each cell in the cross. When the sample departs from proportionality—when some cells are under- or oversampled—it provides less precision than a proportionate sample of the same size. This effect is called the *probability design effect.* Adams et al. (2003) tell us it is most often seen calculated in terms of the analysis weights $w_k = 1/p_k$, where p_k is the probability that individual k is in the sample. We express this effect as an equivalent sample size—the size of a uniformly weighted sample that would yield the same precision as a sample of size n with nonuniform weights:

---

of responses calculated by our method was either strictly smaller than the requirement calculated using the fpc (291 crosses) or at least as large as the requirement calculated *without* the fpc (350 crosses). Dropping the fpc from the calculation of required responses should have no effect on these crosses. Therefore, we do not think our use of the fpc made much of a difference in the design of the sample.

$$ESS = \frac{\left[\sum_k w_k\right]^2}{\sum_k w_k^2}. \quad (4.3)$$

The summation is taken over the n individuals in the sample. It reaches a maximum of n when the weights w_k are all equal—i.e., for a proportionate sample.

All individuals in the same cell will have the same selection probability p_k and hence the same weight w_k. According to Eq. (4.3), the effective sample size is decreased much more by having one or two cells with very low sampling fractions than it is increased by having one or two cells with very high sampling fractions. In other words, it does not help much to oversample a few cells. But it matters a lot if a few cells are undersampled.

Our Fondest Wish

We therefore impose lower limits on the number of people in each cell who are assigned to each marginal cross. These lower limits will be strictly positive unless user requirements dictate otherwise. Initially, we calculate the number of responses we need from each cell to ensure that every cross contains a proportionate sample of the desired size. We calculate the size of a cross as

$$Size(Cross) = \sum_{Cell} Contains(Cross, Cell) \times Pop(Cell). \quad (4.4)$$

In a proportionate sample of the desired size, the number of responses that *Cell* should contribute to *Cross* is

$$PropSam(Cross, Cell) = \frac{Rqmt(Cross)}{Size(Cross)} \times Contains(Cross, Cell) \times Pop(Cell). \quad (4.5)$$

The responses we receive from *Cell* will be distributed over the various assignments. For each assignment, we want to receive enough responses to meet the needs of all crosses with that assignment. Letting *AssignV* stand for any value of the assignment (i.e., *Core*, *A*, *BC*, or *HEALTH*), we write

$$Wish(AssignV, Cell) = \max_{\{Cross \mid Assign(Cross) = AssignV\}} \{PropSam(Cross, Cell)\}. \quad (4.6)$$

Taking stock, *Wish*(*AssignV*,*Cell*) is the number of responses we need from *Cell* for each *AssignV* for our sample to include a proportionate subsample of the desired size in every marginal cross. Many crosses, of course, may include extra responses.

Reducing Overall Demands

Generally, it will not be possible to provide as many as *Wish*(*AssignV*,*Cell*) responses from all cells for all assignments. The CULTURE and HEALTH surveys compete for people in each cell, and within the CULTURE survey, the sections compete as well. Therefore, we have introduced several parameters—earlier, we dubbed them *design levers*—that govern the way we allocate the shortage of responses from a cell among the assignments.

The first design lever specifies the minimum fraction of a proportionate subsample that we are willing to settle for. We define

WishFrac = The minimum fraction of a proportionate sample that must—if at all possible—be included in each marginal cross.

Allocating Shortfalls Between Surveys

Reducing our overall demands may help, but there may be cells from which it is not possible to provide *WishFrac* × *Wish*(*AssignV*,*Cell*) responses to all assignments *AssignV*. For these cells, we specify two more design levers: one to govern how the remaining shortfall is allocated between the CULTURE and HEALTH surveys; the other to govern how any shortfall within the CULTURE survey is allocated among the sections.

We denote by *Hprio* the design lever we use to allocate shortfalls between CULTURE and HEALTH. *Hprio* should be between 0 and 1, with 0 corresponding to absolute priority for the CULTURE survey and 1 corresponding to absolute priority for the HEALTH survey. Giving the HEALTH survey "absolute priority" means assigning the maximum number of individuals possible from *Cell* to the HEALTH survey, up to the number that will yield *WishFrac* × *Wish*("*HEALTH*",*Cell*) responses. The CULTURE survey is left to cope the best it can. Giving the CULTURE survey absolute priority is defined similarly. We denote

Hmax ("*Core*",*Cell*) = The maximum number of responses from *Cell* that can be obtained for the CULTURE survey if the HEALTH survey is given absolute priority.

Hmax("*HEALTH*",*Cell*) = The maximum number of responses from *Cell* that can be obtained for the HEALTH survey if the HEALTH survey is given absolute priority.

Cmax("*Core*",*Cell*) = The maximum number of responses from *Cell* that can be obtained for the CULTURE survey if the CULTURE survey is given absolute priority.

Cmax("*HEALTH*",*Cell*) = The maximum number of responses from *Cell* that can be obtained for the HEALTH survey if the CULTURE survey is given absolute priority.

It is straightforward to calculate the *Hmax* and *Cmax* variables. We have suppressed the index *Cell* in the following equations in the interest of readability, but remember that there is one set of these equations for each *Cell*.

$$Hmax(\text{“}HEALTH\text{”}) = MIN\begin{Bmatrix} SRR(\text{“}HEALTH\text{”}) \times Pop \\ WishFrac \times Wish(\text{“}HEALTH\text{”}) \end{Bmatrix}. \tag{4.7}$$

$$Cmax(\text{“}Core\text{”}) = MIN \left\{ \begin{array}{l} SRR(\text{“}Core\text{”}) \times Pop \\ WishFrac \times MAX \left\{ \begin{array}{l} Wish(\text{“}Core\text{”}) \\ Wish(\text{“}A\text{”}) + 2 \times Wish(\text{“}BC\text{”}) \end{array} \right\} \end{array} \right\}. \quad (4.8)$$

$$Hmax(\text{“}Core\text{”}) = MIN \left\{ \begin{array}{l} Cmax(\text{“}Core\text{”}) \\ \left[\dfrac{\left[\begin{array}{l} ORR(\text{“}Core\text{”}) \times Hmax(\text{“}HEALTH\text{”}) \\ +SRR(\text{“}Core\text{”}) \times \left(SRR(\text{“}HEALTH\text{”}) \times Pop - Hmax(\text{“}HEALTH\text{”}) \right) \end{array} \right]}{SRR(\text{“}HEALTH\text{”})} \right] \end{array} \right\}. \quad (4.9)$$

$$Cmax(\text{“}HEALTH\text{”}) = MIN \left\{ \begin{array}{l} Hmax(\text{“}HEALTH\text{”}) \\ \left[\dfrac{SRR(\text{“}HEALTH\text{”}) \times \left(SRR(\text{“}Core\text{”}) \times Pop - Cmax(\text{“}Core\text{”}) \right)}{SRR(\text{“}Core\text{”}) - ORR(\text{“}Core\text{”})} \right] \end{array} \right\}. \quad (4.10)$$

Then we use the design level *Hprio* to form weighted averages of these two extreme allocations—absolute priority for CULTURE and absolute priority for HEALTH. The results are the minimum responses that we insist each survey receive from *Cell*:

$$\begin{aligned} ResMin(\text{“}Core\text{”}, Cell) = {} & Hprio \times Hmax(\text{“}Core\text{”}, Cell) \\ & + (1 - Hprio) \times Cmax(\text{“}Core\text{”}, Cell). \end{aligned} \quad (4.11)$$

$$\begin{aligned} ResMin(\text{“}HEALTH\text{”}, Cell) = {} & Hprio \times Hmax(\text{“}HEALTH\text{”}, Cell) \\ & + (1 - Hprio) \times Cmax(\text{“}HEALTH\text{”}, Cell. \end{aligned} \quad (4.12)$$

For each cell, we have two more constraints for the linear program:

$$\begin{aligned} & SRR(\text{“}Core\text{”}) \times Sample(\text{“}Core\text{”}, Cell) + ORR(\text{“}Core\text{”}) \times Overlap(\text{“}Core\text{”}, Cell) \\ & \geq ResMin(\text{“}Core\text{”}, Cell). \end{aligned} \quad (4.13)$$

$$SRR(\text{“}HEALTH\text{”}) \times Sample(\text{“}HEALTH\text{”}, Cell) \geq ResMin(\text{“}HEALTH\text{”}, Cell). \quad (4.14)$$

Allocating CULTURE Shortfalls Among Sections

We do this just as we allocated shortfalls between surveys. Denote by *BCprio* the design lever that allocates shortfalls between section A and sections B and C of the CULTURE survey. The lever *BCprio* should be between 0 and 1, with 0 corresponding to absolute priority for section A and 1 corresponding to absolute priority for sections B and C. Giving section A "absolute priority" means assigning the maximum possible number of these responses to section A, up to *WishFrac* × *Wish*("A",*Cell*) responses. Sections B and C receive whatever is left. Giving sections B and C absolute priority is defined similarly. We always divide responses equally between sections B and C. Define:

Amax("A",*Cell*) = The maximum number of responses from *Cell* that can be obtained for section A of the CULTURE survey if it is given absolute priority.

Amax("BC",*Cell*) = The maximum number of responses from *Cell* that can be obtained for sections B and C of the CULTURE survey if section A is given absolute priority.

BCmax("A",*Cell*) = The maximum number of responses from *Cell* that can be obtained for section A of the CULTURE survey if sections B and C are given absolute priority.

BCmax("BC",*Cell*) = The maximum number of responses from *Cell* that can be obtained for sections B and C of the CULTURE survey if they are given absolute priority.

It is straightforward to calculate the *Amax* and *BCmax* variables:

$$Amax(\text{"}A\text{"},Cell) = MIN\begin{Bmatrix} ResMin(\text{"}Core\text{"},Cell) \\ WishFrac \times Wish(\text{"}A\text{"},Cell) \end{Bmatrix}. \quad (4.15)$$

$$BCmax(\text{"}BC\text{"},Cell) = MIN\begin{Bmatrix} \left[\dfrac{ResMin(\text{"}Core\text{"},Cell)}{2}\right] \\ \\ WishFrac \times Wish(\text{"}BC\text{"},Cell) \end{Bmatrix}. \quad (4.16)$$

$$Amax(\text{"}BC\text{"},Cell) = MIN\begin{Bmatrix} BCmax(\text{"}BC\text{"},Cell) \\ \\ \left[\dfrac{ResMin(\text{"}Core\text{"},Cell) - Amax(\text{"}A\text{"},Cell)}{2}\right] \end{Bmatrix}. \quad (4.17)$$

$$BCmax(\text{"}A\text{"},Cell) = MIN\left\{\begin{array}{l} Amax(\text{"}A\text{"},Cell) \\ ResMin(\text{"}Core\text{"}) - 2 \times BCmax(\text{"}BC\text{"},Cell) \end{array}\right\}. \quad (4.18)$$

Then we use the design level *BCprio* to form weighted averages of these two extreme allocations. The results are the minimum responses that we insist each section receive from *Cell*:

$$\begin{aligned} ResMin(\text{"}A\text{"},Cell) = {} & BCprio \times BCmax(\text{"}A\text{"},Cell) \\ & + (1 - BCprio) \times Amax(\text{"}A\text{"},Cell). \end{aligned} \quad (4.19)$$

$$\begin{aligned} ResMin(\text{"}BC\text{"},Cell) = {} & BCprio \times BCmax(\text{"}BC\text{"},Cell) \\ & + (1 - BCprio) \times Amax(\text{"}BC\text{"},Cell). \end{aligned} \quad (4.20)$$

We note that to allocate the shortfall equally among the three sections, *BCprio* should be 2/3, not 1/2.

For each cell we have two more constraints for the linear program:

$$\begin{aligned} & SRR(\text{"}A\text{"}) \times Sample(\text{"}A\text{"},Cell) + ORR(\text{"}BC\text{"}) \times Overlap(\text{"}A\text{"},Cell) \\ & \geq ResMin(\text{"}A\text{"},Cell). \end{aligned} \quad (4.21)$$

$$\begin{aligned} & SRR(\text{"}BC\text{"}) \times Sample(\text{"}BC\text{"},Cell) + ORR(\text{"}BC\text{"}) \times Overlap(\text{"}BC\text{"},Cell) \\ & \geq ResMin(\text{"}BC\text{"},Cell). \end{aligned} \quad (4.22)$$

Equations (4.15) through (4.22) are correct only if *SRR*("*A*") = *SRR*("*BC*") = *SRR*("*Core*") and *ORR*("*A*") = *ORR*("*BC*") = *ORR*("*Core*"). We leave the derivation of more general formulas as an exercise for the reader.

CHAPTER FIVE

Tradeoffs Among Samples of Different Designs

To begin this chapter, we describe the algorithm for generating a sample. Merely because the algorithm uses linear programming—a methodology that solves an optimization problem—to design a sample does not mean that the sample is "optimal" in a real-world sense. It is, of course, optimal in the formal sense that it is a solution to an optimization problem. But the optimization problem that it solves is only an approximation of the real-world problem we actually want to solve.

We therefore need to generate multiple sample designs and compare them in order to choose the one we like best. To generate different samples, we execute the algorithm repeatedly with different values for the design levers described in Chapter Four. To aid in comparing different samples, we define summary measures of sample quality.

The Sample Design Algorithm

As described earlier, we design a sample by solving two linear programs in sequence. The first linear program finds the values of the variables *Sample(Assign,Cell)*, *Overlap(Assign,Cell)*, and *Short(Cross)* that

$$
\begin{aligned}
&\textit{Minimize} && \sum_{\textit{Cross}} \textit{Short}(\textit{Cross}) \\
&\textit{s.t.} && \textit{Eqs.}\ (3.1)-(4.2),(4.13),(4.14),(4.21),(4.22) \\
& && \textit{Sample}(\textit{Assign},\textit{Cell}) \geq 0 \\
& && \textit{Overlap}(\textit{Assign},\textit{Cell}) \geq 0 \\
& && \textit{Short}(\textit{Cross}) \geq 0.
\end{aligned}
\tag{5.1}
$$

This linear program finds a sample for which the total shortfall from desired sample sizes in all the marginal crosses is a minimum.

Now let $Sval(Cross)$ be the value of $Short(Cross)$ obtained by solving Eq. (5.1). The second linear program then finds new values of the variables $Sample(Assign, Cell)$, $Overlap(Assign, Cell)$, and $Short(Cross)$ that

$$\text{Minimize} \quad \sum_{Cell} \begin{pmatrix} Sample(\text{"Core"}, Cell) \\ + Overlap(\text{"Core"}, Cell) \\ + Sample(\text{"HEALTH"}, Cell) \end{pmatrix}$$

$$s.t. \qquad Eqs.\ (3.1) - (4.2), (4.13), (4.14), (4.21), (4.22)$$

$$Short(Cross) \leq Sval(Cross)$$

$$\begin{aligned} & Sample(Assign, Cell) \geq 0 \\ & Overlap(Assign, Cell) \geq 0 \\ & Short(Cross) \geq 0. \end{aligned} \tag{5.2}$$

Equation (5.2) selects a sample from among all the samples that achieve the minimum shortfalls found by the first linear program. The sample it picks is the one that sends out the fewest number of requests for individuals to participate in one of the surveys. Note that if an individual is asked to participate in both surveys (i.e., is in the overlap), he or she is counted twice.

Generating Multiple Samples

To generate different samples, we execute the algorithm repeatedly for different values of the design levers:

- *WishFrac* (minimum fraction of a proportionate sample of the desired size that must—if at all possible—be included in each cross)
- *Hprio* (relative priority given to the HEALTH survey compared with the CULTURE survey)
- *BCprio* (relative priority given to sections B and C of the CULTURE survey compared with section A). Allow overlap (set $ORR(\text{"A"}) = ORR(\text{"BC"}) = ORR(\text{"Core"}) = 0.25$) or do not allow overlap (set $ORR(\text{"A"}) = ORR(\text{"BC"}) = ORR(\text{"Core"}) = 0$) between the CULTURE and HEALTH surveys.

Summary Measures of Sample Quality

We use two types of summary measure, one type for sample sizes and one type for precision. We will be interested in three sample size measures (Table 5.1).

Table 5.1
Summary Sample Size Measures

Measure	Definition
Number of individuals participating in only the CULTURE survey	$Cpart = \sum_{Cell} Sample(\text{"Core"}, Cell)$
Number of individuals participating in both surveys	$Opart = \sum_{Cell} Overlap(\text{"Core"}, Cell)$
Number of individuals participating in the HEALTH survey (includes the overlap)	$Hpart = \sum_{Cell} Sample(\text{"HEALTH"}, Cell)$

For summary measures of precision, we use counts of "problem crosses." A problem cross is a marginal cross whose effective sample size, computed using Eq. (4.3), is smaller than a specified fraction (called the *problem cross criterion*) of the required number of responses (denoted by *Rqmt*(*Cross*)). There are four problem cross measures of interest, one for each possible assignment (i.e., *Core*, *A*, *BC*, and *HEALTH*).

Exploring the Limits

First, we generate samples that give each survey, in turn, absolute priority over the other. We give the HEALTH survey absolute priority by setting *Hprio* = 1, and the CULTURE survey absolute priority by setting *Hprio* = 0. If we also set *WishFrac* = 1, we will generate the samples that provide the greatest precision for each survey group. For these samples, we allow no overlap.

The Best Possible HEALTH Sample

When *Hprio* = 1 and *WishFrac* = 1, our model gives the HEALTH survey all the precision it wants. Every cross with *Assign*(*Cross*) = "HEALTH" has a proportionate subsample with exactly the desired size, so there are no problem crosses for any criterion. The number of individuals participating in the HEALTH survey (*Hpart*) is 36,040.

The Best Possible CULTURE Sample

When *Hprio* = 0 and *WishFrac* = 1, every cross with *Assign*(*Cross*) = "Core" contains a proportionate subsample of the desired size. Many crosses have larger samples than desired, and the larger sample is often not proportionate.

However, there are crosses with *Assign*(*Cross*) = "A" and "BC" that do not contain proportionate subsamples of the desired size. The *BCprio* lever offers a way to trade off the sections against each other. Table 5.2 compares six samples generated for a range of values of *BCprio*.

Note that changes in *BCprio* do not affect the sample size. Changing *BCprio* does affect the total number of problem crosses, using any problem cross criterion. But for any criterion, the total number of problem crosses is close to its minimum when *BCprio* = 2/3. This is also the value of *BCprio* that gives the three sections equal weight. We adopt this value in all subsequent cases.

Table 5.2
Tradeoff Between the CULTURE Survey's Section A and Sections B and C (*Hprio* = 0, *WishFrac* = 1, No Overlap)

BCprio	0	0.333	0.5	0.667	0.8	1
Sample Size Measures						
Cpart	121,861	121,711	121,968	122,212	122,326	122,580
Total Problem Crosses (Sections A + B + C)						
Criterion						
0.2	222	24	0	0	0	150
0.4	236	106	74	75	66	156
0.6	246	134	176	203	221	255
0.8	258	237	251	262	275	281
1.0	260	321	306	320	315	299
Breakout of Problem Crosses for Criterion = 0.8						
Section						
A	0	71	103	120	143	157
B or C	129	83	74	71	66	62

NOTE: To generate the samples discussed in this table, we allow no overlap. But when the CULTURE survey receives absolute priority (*Hprio* = 0), it does not matter whether overlap is allowed.

There is a total of 157 marginal crosses with a section A assignment, and 130 crosses each assigned to sections B and C (see Table 4.1). Table 5.2 shows that it is possible to eliminate all problem crosses in section A, but because sections B and C compete with one another, even setting *BCprio* = 1 leaves half of section B and C crosses as problem crosses at a criterion of 0.8.

We next varied the lever *WishFrac* to see if we could generate a better sample for the CULTURE survey. We anticipated that reducing *WishFrac* would increase the number of problem crosses for large values of the problem cross criterion, but we thought that it might result in fewer problem crosses at lower criteria. That is, there might be more crosses that fail to contain a subsample with all cells equal to or larger than, say, 90 percent of a proportionate sample of the desired size. But this could be compensated by more crosses containing a subsample with all cells equal to or larger than 60 percent of the desired size. This could happen because reducing *WishFrac* makes more samples feasible—i.e., more samples satisfy all the constraints of the linear program. However, the linear programs we have formulated do not minimize the number of problem crosses for any criterion. Only by experimenting can we discover whether changing *WishFrac* produces better samples. Table 5.3 shows the results.

The one consistent effect of reducing *WishFrac* is reducing the sample size.[1] The CULTURE survey team showed no excitement at this result; they judged maintaining precision to be more important than reducing the sample size. Since people were invited to partici-

[1] There is the anomalous effect on the number of problem cells for criterion = 0.4. Almost surely there are multiple solutions to the sequence of linear programs (Eqs. (5.1) and (5.2)). There could well be solutions other than the ones our method happened to generate that would not show this anomalous behavior.

Table 5.3
Effect of *WishFrac* on the CULTURE Survey Sample (*Hprio* = 0, *BCprio* = 2/3, No Overlap)

WishFrac	1.0	0.8	0.6	0.4
		Sample Size		
Cpart	122,212	112,763	104,693	98,928
		Number of Problem Crosses		
Criterion				
0.2	0	0	0	0
0.4	75	70	44	92
0.6	203	207	224	273
0.8	262	279	350	372
1.0	320	374	432	484

pate by email, the cost of the survey was largely independent of sample size (excluding the value of participants' time). Therefore, we set *WishFrac* = 1 in all subsequent designs.

Reconciling CULTURE and HEALTH

It is now time to reconcile the two survey samples with each other. Table 5.4 shows the effect of varying the priority given to the HEALTH survey from 0 (CULTURE receives absolute

Table 5.4
Effect of *Hprio* on the CULTURE and HEALTH Survey Samples (*WishFrac* = 1, *BCprio* = 2/3, No Overlap)

Hprio	0	0.25	0.5	0.75	1
			Sample Sizes		
Cpart	122,212	116,649	111,160	105,731	100,327
Hpart	16,553	21,944	27,138	32,081	36,040
		CULTURE Survey Problem Crosses (out of 715 total)			
Criterion					
0.2	0	0	0	16	335
0.4	75	112	131	163	415
0.6	203	218	233	291	478
0.8	262	270	335	416	515
1.0	320	426	450	476	552
		HEALTH Survey Problem Crosses (out of 190 total)			
Criterion					
0.2	190	0	0	0	0
0.4	190	124	0	0	0
0.6	190	184	79	0	0
0.8	190	190	182	56	0
1.0	190	190	190	190	0

priority) to 1 (HEALTH receives absolute priority). It seemed to the survey team that the CULTURE survey lost a great deal of precision by giving much priority to the HEALTH survey, but the HEALTH survey had rather little precision unless it received a very high priority.

This observation persuaded the survey team to investigate the effect of allowing the two samples to overlap. To recapitulate, participants in the overlap responded to the HEALTH survey at a 50-percent rate (the nominal response rate), but they responded to the CULTURE survey at only a 25-percent rate. Table 5.5 shows the results.

Comparing Tables 5.4 and 5.5, one sees that the CULTURE survey loses much less precision in the samples with overlap than in those without. The precision of the HEALTH survey is hardly affected by the overlap. The total number of participants (*Cpart* + *Hpart*) is hardly affected by either the priority given to the HEALTH survey (*Hprio*) or by whether or not the samples are allowed to overlap, ranging from just under 136,000 to just under 139,000. The total number of surveys distributed, (*Cpart* + *Opart* + *Hpart*), is somewhat more variable, exceeding 163,000 in one sample.

Based on these runs, plus some additional runs not shown here, the survey team decided to construct the CULTURE and HEALTH samples using the design for *WishFrac* = 1, *Hprio* = 0.7, and *BCprio* = 2/3, with overlap permitted.

Table 5.5
Effect of *Hprio* on the CULTURE and HEALTH Survey Samples with Overlap (*WishFrac* = 1, *BCprio* = 2/3)

Hprio	0	0.25	0.5	0.75	1
Sample Sizes					
Cpart	116,419	112,172	107,970	103,887	99,800
Opart	11,790	15,860	19,835	23,656	27,656
Hpart	22,158	26,124	29,919	33,393	36,040
CULTURE Survey Problem Crosses (out of 715 total)					
Criterion					
0.2	0	0	0	0	0
0.4	75	93	112	117	131
0.6	203	207	216	221	228
0.8	262	264	267	281	327
1.0	318	418	423	432	446
HEALTH Survey Problem Crosses (out of 190 total)					
Criterion					
0.2	190	0	0	0	0
0.4	190	90	0	0	0
0.6	190	167	63	0	0
0.8	190	187	160	44	0
1.0	190	190	190	190	0

CHAPTER SIX

Sample Selection

We have now selected a sample design, as specified by the arrays *Sample* (*Assign*,*Cell*) and *Overlap* (*Assign*,*Cell*). In this chapter, we describe how we constructed a list of individuals from the Air Force personnel file (see Chapter Two) who will be invited to take the survey.

As stated previously, our method does not draw a simple random sample of predetermined size from each cell. The presence of small cells makes that method impractical. For example, a person in a cell of size one can be allocated to only one section of one survey, leaving the other surveys/sections with no individuals from that cell. Instead, our method assigns each member of the population to a survey/section with a probability equal to the expected size of the sample from his cell—given by *Sample* (*Assign*,*Cell*) or *Overlap* (*Assign*,*Cell*)—divided by the cell size (given by *Pop* (*Cell*)). The method ensures that the size of the sample actually taken from a large cell is close to the expected sample size for that cell, so if there were no small cells we would very nearly obtain a stratified random sample. The method also ensures that the overall size of the sample is very close to the expected size.

We used a variant of systematic sampling (e.g., see Cochran, 1977, Chapter 8; or Kish, 1995, Chapter 4). In the usual version of systematic sampling, one takes every kth member of the population after a random start. This version selects a proportion ($1/k$) of the population, with each member having the same selection probability.

In our problem, however, members of the population are not simply selected or not selected. Rather, there are eleven possible "bins" to which they can be assigned. Moreover, the probability for each assignment is not the same for all members (though the probability is the same for members in the same cell). Our variant of systematic sampling accommodates these complexities. Figure 6.1 shows the bins to which members can be assigned and the assignment probabilities. In addition, we assign members to bins in stages, and the figure shows the hierarchy of stages we used. It will become clear why we chose to assign members to bins in stages rather than in a single step.

Assigning Members to Bins

As the selection algorithm processes each member of the population in turn, it maintains a running score for each bin. Roughly, the score for a bin represents the difference between the expected number of members assigned to the bin up to that point and the actual number

Figure 6.1
Bins, Their Probabilities, and Their Stages of Assignment

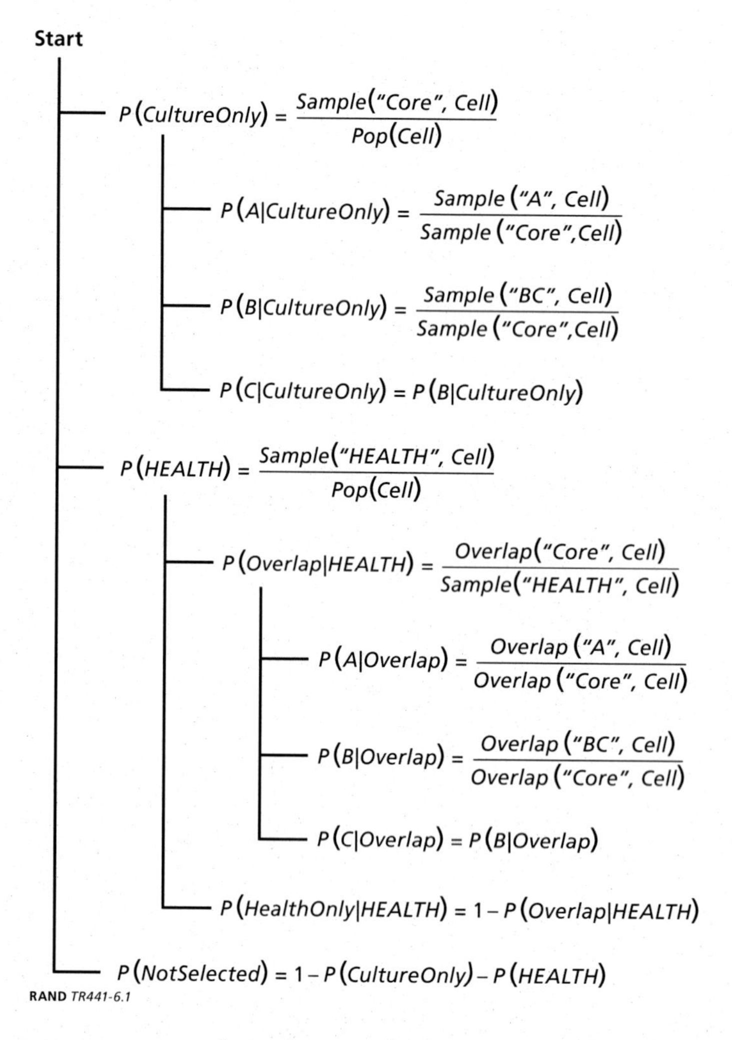

assigned. A negative score means that more people than expected have been assigned to the bin to this point. A positive score means that fewer than expected have been assigned. The bin with the largest score is thus the best candidate to receive the member currently in process.

To illustrate, consider the first stage, which assigns the member to *CultureOnly*, *HEALTH*, or *NotSelected*. We denote the scores of these bins at the time we start processing this member by *Score*(*CultureOnly*), *Score*(*HEALTH*), and *ScoreNotSelected*). The algorithm is as follows:

1. Calculate a test quantity for each bin by adding the probability for each assignment to the current score. That is, *Test*(*CultureOnly*) = *Score*(*CultureOnly*) + *P*(*CultureOnly*), and similarly for the other bins.

2. Put the member in the bin with the largest test quantity, ignoring any bins with zero probabilities.[1] Suppose that this is the *CultureOnly* bin.
3. Calculate the new score for each bin. The new score for the bin that received the member equals the test quantity less 1. The new score for any other bin is equal to its test quantity. Thus, $NewScore(CultureOnly) = Test(CultureOnly) - 1$, while $NewScore(HEALTH) = Test(HEALTH)$.

If the member is assigned to a bin that requires a further stage of assignment, we follow exactly the same procedure but with the probabilities appropriate to the bins now being considered (see Figure 6.1).

Some Facts About Bin Scores

If we set all the scores to 0 before processing the first member of the population, the score a bin has attained after the *N*th member is processed will equal the difference between the expected number and actual number assigned to the bin up to that point. By selecting nonzero starting scores at random, however, we can change the assignments randomly. This is analogous to selecting a random starting point in the usual version of systematic sampling.

However, if a bin's starting score is too large, a string of members from the beginning of the personnel file will be assigned to that bin, even if its probability is small. Conversely, if the score is a large negative number, no members will be assigned until deep into the personnel file, even if its probability is large. The starting scores, therefore, should all be between –1 and +1. It is also useful to insist that the sum of starting scores for the bins involved at each stage be 0. If these conditions are met and we are assigning members to no more than three bins at any stage, we can prove that the scores will all remain between –1 and +1. This need not be true if there are four or more bins at a stage, and this is why we chose to assign members to bins in stages, rather than in a single step.

Theorem 1: The sum of scores for the bins involved at each stage remains 0, no matter how many members have been assigned.

Proof: By assumption, the sum of scores starts at 0. When we assign a member, we add each bin's probability to its score, thus adding 1 to the sum of scores. Then we subtract 1 from one of the scores. The net change in the sum of scores is therefore 0. QED.

Theorem 2: If no more than three bins are involved at a stage, none of their scores can ever drop below –1.

Proof: Before assigning the first member, all scores are at least as large as –1, and the scores for bins involved at each stage sum to 0. Suppose this remains true after assigning member $N-1$. We show that it is still true after assigning member N.

[1] We found that a bin with zero probability for the current member could nevertheless have the highest score. This can happen when the bin in question has had zero probability for a sequence of members. Those members were assigned to other bins, lowering their scores and eventually leaving the score of the bin in question as the highest.

There are three cases: If only one of the three probabilities is positive, its bin is the one to which the member must be assigned. But its new score is its old score plus its probability (which must be 1) minus 1, and hence is unchanged. The scores of the one or two bins with probabilities of 0 are also unchanged.

If exactly two probabilities are positive, then the sum of their two test quantities must be nonnegative. (The sum of the three test quantities equals 1, and the odd test quantity, for the bin with probability 0, cannot be larger than 1.) So the larger of the two "in-play" test quantities is positive. The member is assigned to that test quantity's bin, and the new score, which equals that test quantity minus 1, cannot be smaller than –1.

If all three probabilities are positive, then because the sum of the test quantities is +1, the maximum test quantity must be positive. The member is assigned to that test quantity's bin, and the new score, which equals that test quantity less 1, cannot be smaller than –1. QED.

Theorem 3: If no more than three bins are involved at a stage, none of their scores can ever rise above +1.

Proof: Before assigning the first member, all scores are less than or equal to +1, and the scores for bins involved at each stage sum to 0. Suppose this remains true after assigning member $N-1$. We show that it is still true after assigning member N.

Because the sum of scores after assigning member $N-1$ is 0, the sum of test quantities just before assigning member N is 1. If one of the scores after assigning member N is going to exceed 1, then two test quantities must exceed 1. (It is not possible for any test quantity to exceed 2, since the score from which it was calculated would have to exceed 1.) The third test quantity must therefore be smaller than –1. But this is impossible, since it cannot be smaller than the score from which it was calculated, and that score was at least as large as –1. QED.

The Order of Assigning Members

To obtain the best agreement between the expected sample sizes per cell calculated by the linear program and the number of people in a cell assigned to each bin, we sort the personnel file to put all people in each cell in sequence. However, we want people in large cells to be listed in a random order. And we want the order of very small cells (e.g., cells of containing only one person) to be random. To accomplish these objectives, we randomly assign each person a person identification number (PID). We also assign each cell a cell identification number (CID), constructed as the sum of the cell size *Pop*(*Cell*) plus a random number between 0 and 1. Then we sort the personnel file first by CID from large to small, and within CID by PID.

The Performance of the Selection Algorithm

The fact that the people in a cell are in an unbroken sequence in the file guarantees that the number of people in each bin at the first stage of the assignment hierarchy (see Figure 6.1) is within 2 of the number determined by the linear program. Just before assigning the first member of the cell and just after assigning the last, the number assigned to a bin will be within

1 of the expected number. If the number assigned is high by 1 at the start (end) and low by 1 at the end (start), the number of people assigned to the bin will be 2 less (more) than determined by the linear program. Bins at later stages of the assignment hierarchy can have larger errors. The maximum error for a cell is two per stage.

Since a marginal cross is made up of many cells, the maximum error for a cross can be larger. Figure 6.2 shows the error—i.e., the difference between the responses provided by the sample our algorithm selected minus the responses called for by the linear programming solution—as a function of the responses called for by the linear programming solution. Each data point represents a marginal cross. We use a logarithmic scale for the horizontal axis because the range of cross sizes is so great. Crosses for which the linear program called for less than one response are omitted. The largest absolute difference is less than 15 responses. There are some large percentage differences, but they occur only for crosses from which the linear program took very small samples.

Will the Sample Provide Adequate Precision?

In Chapter Four, we described constraints designed to ensure that meaningful differences between crosses would be statistically significant. For example, suppose we consider it mean-

Figure 6.2
Comparison of Responses per Cross, Sample Versus Linear Program

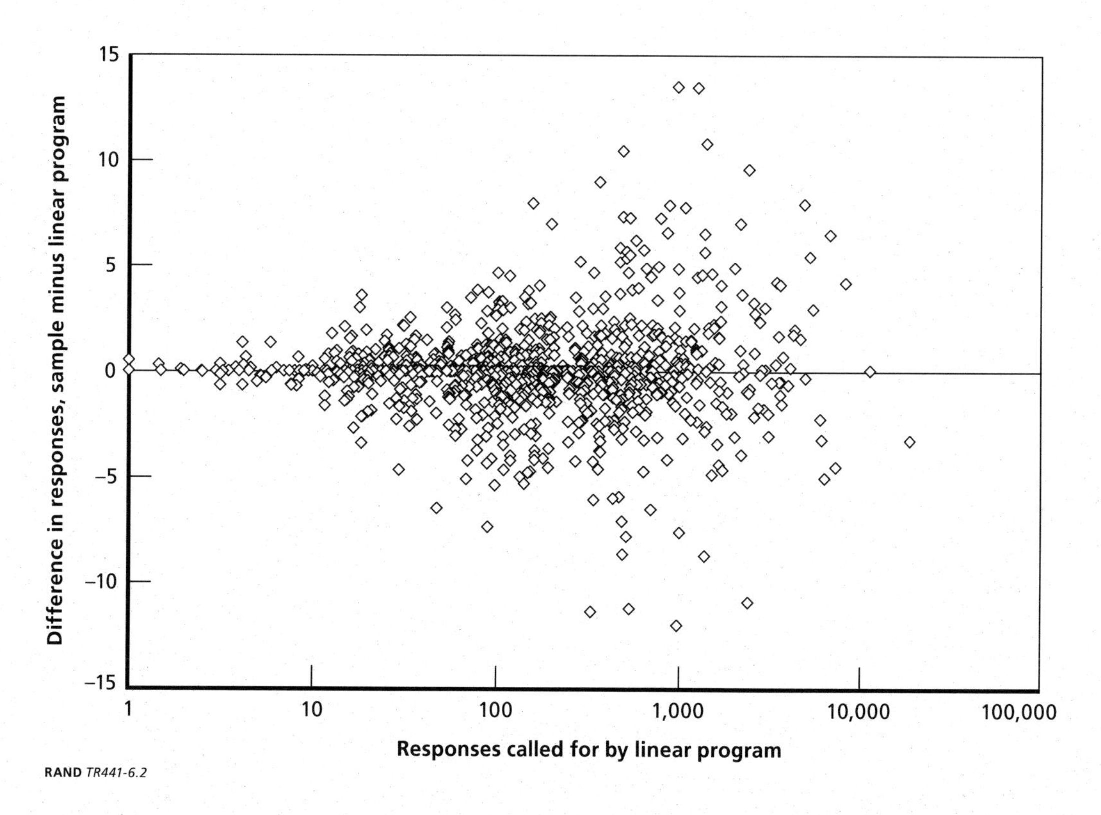

RAND *TR441-6.2*

ingful if vanilla is preferred to chocolate by as many as 5 percent more of the entire population of one cross than of another. Then, upon analyzing the survey responses, we want differences of 5 percent to be statistically significant (usually at a 90- or 95-percent confidence level).

However, the variances we calculate as we analyze the survey responses will depend on factors that were unknown when we designed the sample and fielded the survey. For example, if the actual response rates are lower (higher) than the response rates assumed when we designed the sample, the variances will be higher (lower). Or, if respondents in a cross show broad agreement (disagreement) in their answers to a given question, the estimates of how the entire population of the cross would answer that question will have a small (large) variance.

In addition, we imposed constraints to ensure adequate precision of estimates we anticipated we were going to make. Once we begin analyzing the survey responses, we may discover that some quantities we did not anticipate estimating are much more interesting.

In short, we will inevitably judge the adequacy of the sample using standards that we do not know and hence cannot apply at the time we design the sample. We can only guess what those standards will be—as informed a guess as possible, of course—design the sample to our guess, and let the chips fall where they may.

Bibliography

Adams, John L., Steven L. Wickstrom, Margaret J. Burgess, Paul P. Lee, and José J. Escarce (2003). "Sampling Patients Within Physician Practices and Health Plans: Multistage Cluster Samples in Health Services Research." *Health Services Research,* Vol. 38, No. 6, Part I, pp. 1625–1640.

Chromy, James R. (1987). "Design Optimization with Multiple Objectives." *Proceedings on the Research Methods of the American Statistical Association,* pp. 194–199.

Cochran, William G. (1977). *Sampling Techniques,* 3rd ed. Somerset, N.J.: John Wiley & Sons, Inc.

Elliott, Mark N., Alan M. Zaslavsky, and Paul D. Cleary (2006). "Are Finite Population Corrections Appropriate When Profiling Institutions?" *Health Services & Outcomes Research Methodology,* Vol. 6, pp. 153–156.

GAMS Development Corporation, *General Algebraic Modeling System* (undated). As of March 26, 2007: http://www.gams.com

Hillier, Frederick S., and Gerald J. Lieberman (2005). *Introduction to Operations Research,* 8th ed. New York: McGraw-Hill Higher Education.

Horvitz, D. G., and D. J. Thompson (1952). "A Generalization of Sampling Without Replacement from a Finite Universe." *Journal of the American Statistical Association,* Vol. 47, No. 260, pp. 663–685.

Kish, Leslie (1995). *Survey Sampling.* Wiley Classics Library Edition. Somerset, N.J.: John Wiley & Sons, Inc.

Lu, Wilson, and Randy R. Sitter (2002). "Multi-way Stratification by Linear Programming Made Practical." *Survey Methodology,* Vol. 28, No. 2, pp. 199–207.

Stehman, Stephen V., and W. Scott Overton (1994). "Comparison of Variance Estimators of the Horvith-Thompson Estimator for Randomized Variable Probability Systematic Sampling." *Journal of the American Statistical Association,* Vol. 89, No. 425, pp. 30–43.